Exploring the Fundamentals of Soft Condensed Matter Physics: A Student's Perspective

Vincent Luis

Copyright © [2023]

Title: Exploring the Fundamentals of Soft Condensed Matter Physics: A Student's Perspective

Author's: Vincent Luis

This book was printed and published by [Publisher's: **Vincent Luis**] in [2023]

ISBN:

TABLE OF CONTENT

Chapter 8: Applications of Soft Condensed Matter Physics 64

Chapter 9: Conclusion and Further Reading 73

Chapter 1: Introduction to Soft Condensed Matter Physics

Overview of Soft Condensed Matter Physics

Soft condensed matter physics is a fascinating field that focuses on the study of materials with properties between those of conventional solids and liquids. This subchapter aims to provide students with a comprehensive overview of soft condensed matter physics, exploring its fundamental concepts and applications.

Soft condensed matter refers to a wide range of materials, including polymers, colloids, liquid crystals, and biological systems. These materials are often characterized by their ability to flow and deform under external forces, which makes them highly dynamic and responsive. Understanding their unique behavior and properties is crucial for various fields, including materials science, nanotechnology, biology, and medicine.

One of the key concepts in soft condensed matter physics is the interplay between structure and dynamics. The arrangement of particles and molecules within these materials significantly influences their mechanical, thermal, and electrical properties. By studying the structure, scientists can gain insights into the material's behavior and its response to external stimuli.

Another important aspect of soft condensed matter physics is the study of phase transitions. These transitions occur when a material undergoes a change in its physical properties due to external factors such as temperature, pressure, or concentration. Soft condensed matter systems often exhibit rich phase behavior, with the formation of various ordered and disordered structures. Understanding these phase transitions is crucial for developing new materials with tailored properties.

The subchapter will also delve into the various experimental and theoretical techniques used in soft condensed matter physics. Students will learn about microscopy techniques, such as atomic force microscopy and electron microscopy, which allow for the visualization of materials at the nanoscale. They will also explore spectroscopic techniques, such as X-ray diffraction and nuclear magnetic resonance, which provide insights into the structure and dynamics of soft matter systems.

Furthermore, this subchapter will highlight the interdisciplinary nature of soft condensed matter physics. Students will discover how concepts from physics, chemistry, and biology are combined to understand and manipulate soft materials. They will also explore the practical applications of soft condensed matter physics, such as the development of new drug delivery systems, smart materials, and advanced sensors.

In conclusion, this subchapter provides students with a comprehensive overview of soft condensed matter physics, introducing them to its fundamental concepts, experimental techniques, and practical applications. By exploring the fascinating world of soft matter, students will gain a deeper understanding of the unique behavior and properties of these materials, paving the way for future advancements in condensed matter physics.

Importance of Soft Condensed Matter Physics in Various Fields

The Importance of Soft Condensed Matter Physics in Various Fields

Soft condensed matter physics is a branch of physics that studies the behavior of materials with a complex structure, such as colloids, polymers, liquid crystals, and biological systems. It focuses on understanding the physical properties and processes that occur in these materials at the molecular and mesoscopic scales. This subchapter explores the importance of soft condensed matter physics in various fields and highlights its relevance to students interested in condensed matter physics.

One area where soft condensed matter physics plays a crucial role is in materials science and engineering. By understanding the properties of soft materials, scientists can develop new materials with tailored properties for specific applications. For example, the study of liquid crystals has led to the development of liquid crystal displays (LCDs), which are widely used in electronic devices such as smartphones and televisions. Similarly, the knowledge gained from studying polymers has revolutionized the field of plastics and led to advancements in areas like packaging, medicine, and energy storage.

Soft condensed matter physics also has significant implications in biological systems. Many biological processes, such as protein folding, self-assembly of lipid membranes, and cell mechanics, involve soft materials. By studying these processes, scientists can gain insights into the fundamental principles that govern biological systems and potentially develop new therapeutic strategies for treating diseases. Soft condensed matter physics is also essential in understanding the mechanical properties of tissues and organs, aiding in tissue engineering and regenerative medicine.

Furthermore, soft condensed matter physics plays a crucial role in understanding and developing new technologies. For instance, the study of colloidal suspensions is essential in the development of drug delivery systems, where nanoparticles are designed to carry drugs to specific targets in the body. Soft matter research also contributes to the development of advanced materials for energy applications, such as solar cells and batteries, by investigating the transport of charge and energy in soft materials.

For students interested in condensed matter physics, exploring soft condensed matter physics provides an opportunity to delve into a fascinating and interdisciplinary field. It offers a unique blend of physics, chemistry, materials science, and biology, allowing students to gain a diverse set of skills and knowledge. Moreover, the field offers numerous research opportunities and practical applications, making it an exciting area to pursue further studies or a career.

In summary, soft condensed matter physics plays a pivotal role in various fields, including materials science, biology, and technology. Its study provides insights into the behavior of complex materials and processes, leading to advancements in diverse areas. For students interested in condensed matter physics, exploring soft condensed matter physics offers a rich and rewarding opportunity to contribute to scientific advancements and make a difference in the world.

Scope and Objectives of the Book

Welcome to "Exploring the Fundamentals of Soft Condensed Matter Physics: A Student's Perspective!" This book is specifically designed for students interested in the fascinating field of condensed matter physics, with a particular focus on soft condensed matter. Whether you are a beginner or already have some knowledge of the subject, this book aims to provide a comprehensive understanding of the fundamental concepts and principles underlying soft condensed matter physics.

The scope of this book is to introduce you to the diverse and intriguing world of soft condensed matter, which encompasses a wide range of materials such as polymers, colloids, gels, liquid crystals, and biological systems. Soft condensed matter physics deals with the study of these materials' physical properties, behavior, and interactions on a macroscopic scale, bridging the gap between traditional solid-state physics and liquid-state physics.

Our primary objective is to present the material in a student-friendly manner, keeping in mind the unique needs and perspectives of students like yourselves. We believe that by adopting a student's perspective, we can make the subject more accessible and engaging, enabling you to grasp the underlying concepts and apply them to real-world phenomena.

Throughout the book, we will provide a step-by-step approach to understanding the fundamental principles of soft condensed matter physics. Starting with an introduction to the basic principles of condensed matter physics, we will gradually delve into the specific properties and behavior of soft materials. Topics covered will include the structure and dynamics of polymers, the physics of colloidal systems, the behavior of liquid crystals, and the unique properties of biological systems.

To aid your understanding, we will provide numerous illustrations, diagrams, and real-life examples to make the concepts clearer and more relatable. Additionally, each chapter will conclude with thought-provoking exercises and problem sets designed to reinforce your understanding and encourage critical thinking.

By the end of this book, you will have gained a solid foundation in soft condensed matter physics, enabling you to explore further research opportunities or pursue a career in this exciting field. We hope that this book serves as a valuable resource and companion throughout your journey in understanding and exploring the fundamentals of soft condensed matter physics.

Chapter 2: Basic Concepts in Soft Condensed Matter Physics

Molecular Structure and Interactions

Understanding the molecular structure and interactions is fundamental to unraveling the mysteries of soft condensed matter physics. In this subchapter, we delve into the fascinating world of molecules and how they interact with each other, forming the basis of various soft materials that surround us.

The foundation of soft condensed matter physics lies in the study of materials that possess structural order on a molecular scale, yet exhibit fluid-like properties. This subchapter aims to provide students with a comprehensive overview of the molecular structure and interactions that govern the behavior of soft materials.

We start by exploring the concept of molecular structure, emphasizing the importance of understanding the arrangement of atoms within a molecule. We discuss the different types of bonds that hold atoms together, including covalent, ionic, and hydrogen bonds. Through illustrative examples, we demonstrate how variations in molecular structure can lead to distinct properties and behaviors of soft materials.

Next, we delve into the interactions between molecules, which are crucial for understanding the collective behavior of soft materials. Van der Waals forces, electrostatic interactions, and hydrophobic interactions are among the key forces we explore. We explain how these interactions influence the stability, phase transitions, and self-assembly of soft materials, shedding light on their intriguing properties.

Furthermore, we introduce students to the concept of intermolecular forces and their role in determining the macroscopic properties of soft materials. We explain how intermolecular forces can give rise to phenomena such as surface tension, viscosity, and elasticity, which are integral to understanding the behavior of soft condensed matter.

To provide a comprehensive perspective, we also discuss the importance of molecular simulations and experimental techniques in studying molecular structure and interactions. We highlight the significance of computational tools, such as molecular dynamics simulations and quantum mechanics calculations, in unraveling the behavior of soft materials at the molecular level.

Throughout the subchapter, we present real-world examples and case studies that demonstrate the relevance of molecular structure and interactions in a variety of soft condensed matter systems. By the end, students will gain a solid foundation in understanding the molecular underpinnings of soft materials, enabling them to explore the vast and exciting field of condensed matter physics with confidence.

In conclusion, "Molecular Structure and Interactions" is a crucial subchapter in our book, "Exploring the Fundamentals of Soft Condensed Matter Physics: A Student's Perspective." It equips students with the necessary knowledge to comprehend the behavior of soft materials at the molecular level. By understanding the intricacies of molecular structure and interactions, students will be able to unlock the secrets of soft condensed matter physics and contribute to advancements in this captivating field.

Thermodynamics of Soft Condensed Matter Systems

In the fascinating realm of condensed matter physics, the study of soft condensed matter systems holds a special place. Soft condensed matter refers to materials that exhibit both solid-like and liquid-like properties, such as polymers, colloids, and liquid crystals. Understanding the thermodynamics of these systems is crucial for unraveling the fundamental principles that govern their behavior and for designing innovative materials with tailored properties.

This subchapter delves into the intricacies of the thermodynamics of soft condensed matter systems, providing students with a comprehensive overview of the key concepts and principles involved.

To begin, we explore the concept of equilibrium and its relevance in soft condensed matter systems. Equilibrium is a state in which the material's macroscopic properties no longer change with time. By studying the equilibrium conditions, students will gain insights into the stability and phase behavior of soft materials. We discuss the role of free energy, a fundamental thermodynamic quantity, and its connection to the equilibrium state.

Next, we delve into the phase transitions within soft condensed matter systems. Phase transitions occur when materials undergo a sudden change in their properties as a result of external factors like temperature or pressure. Students will learn about the different types of phase transitions, such as liquid-gas transitions and crystallization, and the associated thermodynamic quantities like entropy and enthalpy.

Furthermore, this subchapter explores the role of entropy in soft condensed matter systems. Entropy is a measure of the disorder or randomness within a system and plays a crucial role in determining the equilibrium configurations and phase behavior of soft materials. Students will gain a deep understanding of how entropy drives phase transitions and affects the overall thermodynamics of these systems.

Lastly, we discuss the thermodynamics of specific soft condensed matter systems, such as polymers and colloidal suspensions. Students will learn about the unique properties and behavior of these materials, including the role of molecular interactions and the influence of external factors on their thermodynamics.

Through clear explanations, illustrative examples, and engaging exercises, this subchapter aims to equip students with a solid foundation in the thermodynamics of soft condensed matter systems. By grasping these fundamental concepts, students will be empowered to explore the vast possibilities and applications of soft condensed matter physics, from designing new materials to understanding complex biological systems.

Statistical Mechanics and Soft Matter

In the fascinating realm of condensed matter physics, one of the most captivating and relevant subfields is the study of soft matter. Soft matter refers to a wide range of materials, including polymers, colloids, liquid crystals, and biological macromolecules, which exhibit unique physical properties due to their flexible and complex structures. Understanding the behavior of soft matter is crucial in various scientific disciplines, including materials science, chemistry, and biology.

This subchapter delves into the captivating intersection between statistical mechanics and soft matter, shedding light on the fundamental principles and concepts that underpin the behavior of these intriguing materials.

Statistical mechanics provides a powerful framework for describing and predicting the properties of large ensembles of particles, such as molecules in a fluid or atoms in a solid. By considering the statistical behavior of individual particles, statistical mechanics enables us to connect the microscopic details to macroscopic observables, offering insights into the collective behavior of soft matter systems.

The subchapter begins by introducing the key concepts of statistical mechanics, including the fundamental postulates of statistical mechanics, such as the ergodic hypothesis and the principle of equal a priori probabilities. These principles lay the foundation for understanding the statistical behavior of soft matter systems and how it relates to their thermodynamic properties.

Next, the subchapter explores the statistical mechanics of polymers, which are essential components of many soft matter systems. Polymers, such as DNA and synthetic polymers, exhibit fascinating behavior due to their long, flexible chains. The subchapter covers topics such as polymer statistics, polymer conformational transitions, and the statistical mechanics of polymer solutions, providing students with a comprehensive understanding of polymer physics.

Moving forward, the subchapter delves into the statistical mechanics of colloids, which are particles suspended in a fluid medium. Colloidal systems exhibit rich phase behavior due to the interplay between particle interactions, thermal fluctuations, and external fields. The subchapter discusses key topics such as colloid phase diagrams, self-assembly, and the statistical mechanics of colloidal suspensions.

Lastly, the subchapter touches upon the statistical mechanics of liquid crystals, which are materials that exhibit both fluid-like and solid-like properties. Liquid crystals are of immense technological importance due to their applications in displays and sensors. The subchapter explores the statistical mechanics of liquid crystals, including the molecular origins of liquid crystal phases and the statistical mechanics of liquid crystal defects.

By combining the principles of statistical mechanics with the fascinating world of soft matter, this subchapter equips students with the tools and knowledge to understand and explore the fundamental behavior of these materials. Through clear explanations, illustrative examples, and exercises, students will gain a solid foundation in statistical mechanics and its applications to soft condensed matter physics, empowering them to further pursue research and advancements in this captivating field.

Phase Transitions in Soft Condensed Matter

In the fascinating world of soft condensed matter physics, the study of phase transitions plays a crucial role in understanding the behavior and properties of various materials. Phase transitions refer to the transformations that occur in a system when certain external conditions, such as temperature or pressure, cross a critical point. These transitions are characterized by abrupt changes in the system's properties, such as its structure, symmetry, or response to external stimuli.

In this subchapter, we will delve into the intriguing realm of phase transitions in soft condensed matter, focusing on their significance and the fundamental concepts associated with them. This knowledge is essential for students seeking a deeper understanding of condensed matter physics.

Firstly, we will explore the different types of soft condensed matter systems in which phase transitions take place. Soft condensed matter refers to materials that are neither purely solid nor completely liquid, encompassing a wide range of substances such as polymers, colloids, liquid crystals, and biological macromolecules. By understanding the nature of these systems, we can appreciate the unique characteristics and behaviors that arise during phase transitions.

Next, we will discuss the role of order parameters in characterizing phase transitions. Order parameters are variables that undergo changes during a phase transition, serving as indicators of the system's transformation. Examples of order parameters include density, magnetization, and orientational order. By studying the behavior of these parameters, we can identify different phases and gain insights into the underlying physics of the system.

We will then delve into the various types of phase transitions that occur in soft condensed matter. These include continuous or second-order phase transitions, where there is a gradual change in the order parameter, and discontinuous or first-order phase transitions, characterized by a sudden jump in the order parameter. The study of critical phenomena and universality classes will also be explored, highlighting the universal behavior and scaling laws that govern phase transitions.

Finally, we will discuss the practical applications of phase transitions in soft condensed matter physics. From the development of new materials with tailored properties to the design of advanced technologies, the understanding of phase transitions has wide-ranging implications across various fields of science and engineering.

By the end of this subchapter, students will have gained a comprehensive understanding of phase transitions in soft condensed matter and their significance in condensed matter physics. Armed with this knowledge, they will be equipped to explore further research and contribute to the exciting advancements in this field.

Chapter 3: Rheology and Mechanical Properties of Soft Condensed Matter

Viscoelasticity

Viscoelasticity is a fascinating property of materials that combines the characteristics of both viscosity and elasticity. In this subchapter, we will explore the fundamentals of viscoelasticity, its significance in condensed matter physics, and its applications in various fields.

Viscoelastic materials, such as gels, polymers, and biological tissues, exhibit a time-dependent response to applied stress or strain. Unlike purely elastic materials that instantly regain their original shape after deformation, viscoelastic materials display a delayed recovery and may even exhibit permanent deformation. This behavior arises from the internal friction within the material, which allows it to dissipate energy and exhibit both solid-like and fluid-like properties.

Understanding viscoelasticity is crucial in condensed matter physics as it plays a vital role in many natural and synthetic materials. For instance, the behavior of viscoelastic materials is essential in fields like biomechanics, where it helps us comprehend the mechanical properties of tissues and organs. Moreover, viscoelasticity has significant implications in the development of advanced materials, such as smart polymers, shape-memory alloys, and self-healing materials.

The study of viscoelasticity involves various theoretical models and experimental techniques. One of the foundational models is the Maxwell model, which describes viscoelastic behavior using a combination of a spring and a dashpot. This model allows us to understand the time-dependent response of materials under different loading conditions. Another widely used model is the Kelvin-Voigt model, which describes how viscoelastic materials respond to a sudden applied stress.

To characterize viscoelastic materials, researchers employ techniques such as rheology, which measures the flow and deformation properties of materials under varying conditions. This allows us to determine parameters like viscosity, elasticity, and relaxation times, which are crucial in understanding the viscoelastic behavior of different materials.

Furthermore, studying the viscoelasticity of materials helps us comprehend various phenomena, including stress relaxation, creep, and hysteresis. These phenomena find applications in fields like materials science, engineering, and even in the study of earthquakes and seismic waves.

In conclusion, viscoelasticity is a fundamental property of soft condensed matter that combines the characteristics of both viscosity and elasticity. Understanding the behavior of viscoelastic materials is crucial in condensed matter physics, as it has significant implications in fields like biomechanics and material science. By studying the theoretical models, experimental techniques, and characterization methods associated with viscoelasticity, students gain a comprehensive understanding of this intriguing property and its applications in various niches of condensed matter physics.

Flow Behavior and Viscosity

In the fascinating realm of soft condensed matter physics, understanding the flow behavior and viscosity of materials is of utmost importance. Whether you're studying the properties of polymers, colloids, liquid crystals, or biological systems, delving into the intricacies of flow can provide valuable insights into the behavior of these materials.

Flow behavior refers to how substances respond when subjected to external forces, such as shear stress. When a force is applied to a material, it may either flow smoothly or exhibit a more complex behavior. The study of flow behavior encompasses various phenomena such as shear thinning, shear thickening, and yield stress. Exploring these phenomena allows us to comprehend the intricate interplay between the particles, their interactions, and the applied forces.

Viscosity, a fundamental property of fluids, plays a crucial role in flow behavior. It characterizes a fluid's resistance to flow and is often described as the thickness or stickiness of a material. The measurement of viscosity allows us to quantify the ease with which a fluid flows, and it can vary significantly depending on the composition and structure of the material.

For students delving into condensed matter physics, understanding the factors influencing viscosity is paramount. The molecular structure, temperature, and concentration of solutes all impact viscosity. By examining these factors, we can gain valuable insights into the behavior of complex materials and their applications in various fields.

Moreover, studying flow behavior and viscosity enables us to explore a wide range of phenomena, such as the flow of liquid crystals in display technology, the behavior of biological fluids, and the rheology of polymers. Connecting these concepts to real-world applications allows students to appreciate the relevance and implications of their studies.

To delve deeper into the world of flow behavior and viscosity, we will explore various experimental techniques used to measure viscosity accurately. These techniques include viscometry, rheometry, and optical methods. By employing these tools, students can quantify and analyze the behavior of soft condensed matter materials, providing further insights into their properties and behavior.

In conclusion, the subchapter on flow behavior and viscosity provides students of condensed matter physics with a solid foundation to understand the complex dynamics of soft materials. By grasping the principles governing flow behavior and viscosity, students can unlock the secrets of various materials and their applications in fields ranging from materials science to biophysics.

Rheological Models and Measurements

Rheology is the study of the flow of matter, particularly in soft condensed matter systems such as liquids, colloids, and polymers. Understanding the rheological properties of these materials is crucial in various fields, including condensed matter physics. In this subchapter, we will explore the fundamental concepts and techniques used in rheological models and measurements.

One of the key aspects of rheology is the characterization of material behavior under different flow conditions. Rheological models provide a mathematical framework to describe and predict the flow properties of soft condensed matter. These models are based on fundamental principles and can be used to understand and analyze complex flows.

There are several commonly used rheological models, including the Newtonian model, the power-law model, and the viscoelastic models. The Newtonian model describes materials that exhibit a linear relationship between the applied stress and the resulting deformation rate. Many simple liquids, such as water, follow this model. The power-law model, on the other hand, is used to describe non-Newtonian fluids, which exhibit a non-linear relationship between stress and deformation. Examples include paints, gels, and certain food products. Viscoelastic models are used for materials that exhibit both viscous (flow-like) and elastic (solid-like) behavior, such as polymers. These models are essential in understanding the unique properties of these materials, including their ability to store and release energy.

To validate and refine these rheological models, accurate measurements are required. Rheological measurements involve applying controlled stress or strain to a material and measuring its response. Various instruments, such as rheometers and viscometers, are used to perform these measurements. Rheometers allow for precise control of the applied stress or strain, while viscometers measure the viscosity of a material under specific flow conditions.

The subchapter will delve into the different techniques used in rheological measurements, including steady-state and oscillatory measurements. Steady-state measurements involve applying a constant stress or strain and measuring the resulting flow properties, such as viscosity or shear rate. Oscillatory measurements, on the other hand, involve applying an oscillating stress or strain and measuring the material's response. These measurements provide valuable information about the material's viscoelastic properties, including storage and loss moduli.

Understanding rheological models and measurements is crucial for students in the field of condensed matter physics. It provides a foundation for studying the behavior of soft condensed matter and its applications in various industries, such as food, cosmetics, and materials science. By grasping the concepts and techniques discussed in this subchapter, students can gain a comprehensive understanding of rheology and its significance in studying soft condensed matter physics.

Mechanical Properties of Soft Condensed Matter Systems

Soft condensed matter physics deals with the study of materials that are neither completely solid nor completely liquid, but rather something in between. These materials possess unique mechanical properties that make them fascinating subjects of study. In this subchapter, we will explore the various mechanical properties of soft condensed matter systems and delve into their significance in the field of condensed matter physics.

One of the fundamental mechanical properties of soft condensed matter systems is elasticity. Unlike traditional solids, which exhibit linear elasticity, soft materials have a nonlinear response to external forces. This nonlinearity arises from the complex interplay between the material's structure, interactions, and dynamics. Understanding the elastic behavior of soft matter is crucial in fields such as biophysics, where the mechanical properties of tissues and cells play a vital role in their functions.

Another important aspect of soft condensed matter is its ability to flow. Unlike traditional liquids, soft matter exhibits viscoelastic behavior, meaning it possesses both viscous and elastic properties. This unique characteristic allows these materials to flow under stress while also retaining some degree of elasticity. The study of flow in soft matter has far-reaching implications, ranging from understanding the behavior of polymers and gels to developing better drug delivery systems.

Soft materials also display interesting rheological properties. Rheology is the study of how materials deform and flow under the influence of external forces. Understanding the rheology of soft condensed matter is crucial for various industrial applications, such as the design of paints, adhesives, and food products. Additionally, the rheological behavior of soft materials has implications in fields like geophysics, where understanding the flow of complex fluids in porous media is essential.

Furthermore, this subchapter will explore the mechanical response of soft matter to external stimuli, such as temperature, pressure, and electric fields. Soft condensed matter systems exhibit phase transitions, where their mechanical properties undergo abrupt changes as these stimuli are varied. Understanding these transitions is essential for designing new materials with tailored properties and for studying the behavior of complex systems, such as liquid crystals and colloids.

In conclusion, the mechanical properties of soft condensed matter systems are at the heart of condensed matter physics. Their unique behaviors, such as nonlinear elasticity, viscoelasticity, and complex rheology, make them intriguing subjects of study. As students exploring the fundamentals of soft condensed matter physics, understanding these mechanical properties will provide a solid foundation for further research and applications in a wide range of fields, from biophysics to materials science and beyond.

Chapter 4: Self-Assembly and Supramolecular Structures

Self-Assembly Phenomena in Soft Condensed Matter Systems

In the realm of Condensed Matter Physics, one of the most fascinating aspects is the study of self-assembly phenomena in soft condensed matter systems. These systems, which encompass a wide range of materials such as polymers, colloids, and liquid crystals, have garnered significant attention due to their ability to spontaneously organize into intricate and functional structures.

Self-assembly refers to the process by which individual components, driven by fundamental physical forces, come together to form ordered arrangements without external intervention. This phenomenon is particularly prominent in soft condensed matter systems due to the interplay between thermal fluctuations and the inherent interactions between the constituent particles.

The subchapter "Self-Assembly Phenomena in Soft Condensed Matter Systems" aims to provide students with a comprehensive understanding of this intriguing field of study. It begins by introducing the fundamental concepts of self-assembly, emphasizing the role of entropy, enthalpy, and intermolecular forces in driving the organization of soft matter systems.

Further exploration delves into the different types of self-assembly observed in soft condensed matter, including micelle formation, liquid crystal phase transitions, and the assembly of colloidal particles. The subchapter highlights the distinct characteristics and mechanisms behind each type, shedding light on the underlying principles governing their formation.

Moreover, the subchapter explores the various techniques employed to study self-assembly phenomena, such as microscopy, spectroscopy, and scattering techniques. It also delves into computational methods used to model and simulate these systems, enabling students to gain a deeper understanding of the intricate dynamics involved in self-assembly.

To illustrate real-world applications, the subchapter showcases examples where self-assembly has been harnessed to create functional materials with tailored properties. These applications range from drug delivery systems and nanotechnology to the fabrication of advanced materials for electronic devices and energy storage.

Throughout the subchapter, emphasis is placed on providing a student-friendly perspective, ensuring that complex concepts are explained in a clear and accessible manner. The content is accompanied by illustrative figures, diagrams, and real-world examples to enhance the learning experience.

By immersing themselves in the world of self-assembly phenomena in soft condensed matter systems, students will gain a deep appreciation for the elegance and complexity of this field. They will develop a solid foundation to explore and contribute to the ongoing research and development in Condensed Matter Physics, paving the way for future breakthroughs in the design and fabrication of advanced materials.

Molecular Aggregates and Micelles

In the world of soft condensed matter physics, molecular aggregates and micelles play a crucial role. These fascinating structures form when molecules come together to self-assemble, creating larger entities with unique properties and behaviors. In this subchapter, we will explore the fundamentals of molecular aggregates and micelles, providing students in the field of condensed matter physics with a comprehensive understanding of their formation and significance.

Molecular aggregates are formed when individual molecules interact with each other through weak forces, such as van der Waals interactions or hydrogen bonding. As a result of these interactions, the molecules cluster together to minimize their overall energy. This clustering leads to the formation of larger structures that often exhibit distinct properties from the individual molecules themselves. For example, molecular aggregates can display altered optical, electrical, or mechanical properties, making them useful in various applications ranging from drug delivery systems to optoelectronics.

One particular type of molecular aggregate that is frequently encountered is the micelle. Micelles are formed when amphiphilic molecules, which have both hydrophilic (water-loving) and hydrophobic (water-repelling) parts, self-assemble in a solvent. The hydrophilic segments face the surrounding solvent, while the hydrophobic segments cluster together in the core, creating a spherical structure. Micelles can be found in numerous everyday products, such as soaps and detergents, where they help solubilize hydrophobic substances in water.

Understanding the formation and behavior of molecular aggregates and micelles is crucial for studying a wide range of phenomena in soft condensed matter physics. The self-assembly processes of these structures are influenced by factors such as temperature, concentration, and molecular properties. By manipulating these parameters, scientists can control the size, shape, and stability of molecular aggregates and micelles, allowing them to tailor their properties for specific applications.

Throughout this subchapter, we will delve into the theoretical framework and experimental techniques used to study molecular aggregates and micelles. We will explore the thermodynamics behind their formation, the various types of molecular interactions involved, and the role of kinetics in their assembly. Additionally, we will discuss advanced characterization techniques, such as microscopy and spectroscopy, that enable the visualization and analysis of these structures at the molecular level.

By the end of this subchapter, students will have gained a solid foundation in the principles of molecular aggregates and micelles, equipping them with the necessary knowledge to delve deeper into the fascinating world of soft condensed matter physics. Whether pursuing further research or applying their understanding to practical applications, students will be well-prepared to contribute to the exciting advancements in this field.

Liquid Crystals and Liquid Crystal Polymers

One of the most fascinating and rapidly evolving areas in the field of Condensed Matter Physics is the study of liquid crystals and liquid crystal polymers. These materials exhibit unique properties that make them extremely useful in various technological applications, ranging from displays in electronic devices to drug delivery systems. In this subchapter, we will explore the fundamentals of liquid crystals and liquid crystal polymers, providing an overview of their structure, behavior, and applications.

Liquid crystals are a state of matter that exists between a solid and a liquid. Unlike conventional liquids, liquid crystals possess an ordered molecular arrangement, similar to that of a solid, while still maintaining some of the fluidity of a liquid. This unique combination of properties gives rise to their exceptional optical and electrical characteristics. Liquid crystals can be further classified into several types based on their molecular arrangement, including nematic, smectic, and cholesteric phases.

The behavior of liquid crystals is strongly influenced by external factors such as temperature, electric fields, and pressure. By manipulating these parameters, we can control the orientation and alignment of the liquid crystal molecules, leading to remarkable changes in their optical properties. This phenomenon, known as the electro-optic effect, forms the basis of liquid crystal displays (LCDs) found in various electronic devices such as televisions, smartphones, and computer monitors.

Liquid crystal polymers, on the other hand, are long-chain macromolecules that exhibit both liquid crystal and polymer properties. These materials possess a unique combination of mechanical strength, flexibility, and self-organizing behavior, making them ideal for applications such as smart materials, sensors, and drug delivery systems. Liquid crystal polymers can form complex hierarchical structures, allowing for precise control over their mechanical and optical properties.

In recent years, researchers have been exploring the potential of liquid crystals and liquid crystal polymers in emerging fields such as photonics, nanotechnology, and biomedicine. By tailoring the chemical structure and composition of these materials, scientists can create advanced materials with tunable properties to suit specific applications.

In conclusion, the study of liquid crystals and liquid crystal polymers represents an exciting and dynamic field within Condensed Matter Physics. These materials offer a wealth of opportunities for technological advancements and scientific exploration. By understanding their unique properties and harnessing their potential, students can contribute to the development of innovative solutions in various disciplines, ranging from electronics to medicine.

Colloidal Particles and Nanoparticles

In the fascinating world of soft condensed matter physics, colloidal particles and nanoparticles hold a special place. These tiny entities, which range in size from nanometers to micrometers, exhibit unique properties that make them of utmost importance in various scientific and technological applications. In this subchapter, we will delve into the fundamentals of colloidal particles and nanoparticles, exploring their characteristics, behavior, and the exciting possibilities they offer within the realm of condensed matter physics.

Colloidal particles are suspended in a medium, such as a liquid or gas, and are typically composed of macromolecules or small particles. Their size and interactions with the surrounding medium play a crucial role in determining their behavior. Due to their small size, colloidal particles experience Brownian motion, constantly jiggling and colliding with other particles. This motion is influenced by factors such as temperature and viscosity, providing insights into the particle dynamics and the properties of the surrounding medium.

Nanoparticles, on the other hand, are even smaller than colloidal particles, typically measuring less than 100 nanometers. Their size places them in the quantum regime, where quantum mechanics governs their behavior. This leads to intriguing phenomena, such as quantum confinement and surface effects, which can drastically alter the properties of nanoparticles compared to their bulk counterparts. Understanding these effects is crucial for unlocking the potential of nanoparticles in fields ranging from medicine to electronics.

The study of colloidal particles and nanoparticles has significant implications in various disciplines. In materials science, researchers harness the unique properties of nanoparticles to design innovative materials with tailored characteristics, such as enhanced strength or conductivity. In biology and medicine, colloidal particles find applications in drug delivery systems and imaging techniques, revolutionizing the way we diagnose and treat diseases. In environmental science, they play a role in pollution control and remediation efforts.

Moreover, the manipulation and control of colloidal particles and nanoparticles have paved the way for the development of new technologies. Scientists are exploring techniques to assemble and organize these particles into complex structures, leading to advancements in nanofabrication and nanotechnology. Additionally, advances in imaging and characterization methods allow us to observe and understand the behavior of individual particles, providing invaluable insights into their properties and interactions.

In conclusion, colloidal particles and nanoparticles are fascinating entities that hold immense potential in the field of condensed matter physics. Their unique properties and behavior unlock a myriad of possibilities in various scientific and technological domains. By studying and harnessing these particles, we can push the boundaries of knowledge and develop innovative solutions to address the challenges of our modern world.

Chapter 5: Soft Matter in Biological Systems

Biological Macromolecules

In the fascinating world of soft condensed matter physics, one area of study that holds immense significance is the exploration of biological macromolecules. These molecules, which are essential for life, play a crucial role in various biological processes and are at the heart of understanding the principles of condensed matter physics.

Biological macromolecules are large complex molecules that are found in all living organisms. They are composed of smaller units called monomers, which are linked together through chemical bonds. The four major classes of biological macromolecules are proteins, nucleic acids, carbohydrates, and lipids.

Proteins, often referred to as the workhorses of the cell, are involved in a wide range of functions such as catalyzing chemical reactions, regulating gene expression, and providing structural support. Their structure and function are intricately linked, and understanding the principles governing their folding and dynamics is of utmost importance. Soft condensed matter physics provides valuable insights into the behavior of proteins, helping us comprehend how they fold into their three-dimensional structure and how they interact with other molecules.

Nucleic acids, including DNA and RNA, are responsible for storing and transmitting genetic information. These macromolecules have unique properties that allow them to form double helices, which serve as the blueprint for life. Soft condensed matter physics aids in unraveling the mysteries of DNA and RNA, shedding light on their intricate folding patterns and providing tools to manipulate and study them.

Carbohydrates, or sugars, are not only a vital source of energy but also play crucial roles in cell-cell communication and recognition. Soft condensed matter physics investigates the behavior of carbohydrates in solution, revealing their interactions with other macromolecules and their role in biological processes such as cell adhesion and signaling.

Lipids, on the other hand, form the building blocks of cell membranes and are essential for maintaining the integrity of cells. Soft condensed matter physics offers insights into the self-assembly properties of lipids, understanding their behavior in different environments, and their implications in various cellular processes.

By delving into the study of biological macromolecules, students of condensed matter physics gain a deeper understanding of the fundamental principles that govern life itself. The combination of physics and biology provides a unique perspective, allowing us to explore and manipulate these molecules in ways that can have profound implications in fields such as medicine, biotechnology, and materials science.

In conclusion, the exploration of biological macromolecules is a captivating subfield of condensed matter physics. By studying the structure, dynamics, and interactions of proteins, nucleic acids, carbohydrates, and lipids, students can gain a comprehensive understanding of the fundamental principles underlying life and contribute to the advancement of various scientific disciplines.

Biomembranes and Lipid Bilayers

In the realm of soft condensed matter physics, the study of biomembranes and lipid bilayers holds immense significance. These structures, which form the fundamental components of living cells, are crucial for maintaining the integrity and functionality of biological systems. Understanding their properties is essential for comprehending various biological processes and developing new insights into the workings of life itself.

Biomembranes are thin, flexible sheets composed of lipids, proteins, and other molecules. They serve as barriers, separating the interior of cells from the external environment, regulating the passage of ions and molecules, and facilitating communication between cells. The lipid bilayer, a primary structural motif of biomembranes, consists of two layers of lipid molecules with their hydrophilic (water-loving) heads facing outwards and their hydrophobic (water-repelling) tails facing inward, creating a stable and impermeable barrier.

One of the fascinating aspects of biomembranes is their self-assembly capability. Lipid molecules, due to their amphiphilic nature, spontaneously organize themselves into bilayers when placed in an aqueous environment. This self-assembly process is driven by the hydrophobic effect, which minimizes the contact between the hydrophobic tails and water molecules. Through this process, lipid bilayers adopt a variety of structures, including planar membranes, vesicles, micelles, and liposomes, each with its own unique properties and functions.

Studying biomembranes and lipid bilayers involves investigating their mechanical and thermodynamic properties. Condensed matter physicists employ various experimental and theoretical techniques to understand the structural dynamics, elasticity, and phase behavior of these systems. For example, techniques like X-ray and neutron scattering, nuclear magnetic resonance (NMR), and electron microscopy provide insights into the organization and arrangement of lipid molecules within bilayers. Molecular dynamics simulations and statistical mechanics approaches are used to elucidate the thermodynamic behavior and phase transitions exhibited by these systems.

Moreover, the study of biomembranes and lipid bilayers has important implications in the fields of biotechnology and drug delivery. Lipid-based nanoparticles and liposomes can be engineered to encapsulate drugs and deliver them to specific targets, offering a potential solution for targeted therapy. Additionally, understanding the behavior of biomembranes is critical in the development of artificial organs, bioengineered tissues, and drug screening platforms.

In conclusion, the exploration of biomembranes and lipid bilayers is a captivating and multidisciplinary field within condensed matter physics. By delving into the fundamental properties and behaviors of these structures, we gain valuable insights into the workings of living systems and open doors to a wide range of applications in biotechnology. As students of condensed matter physics, the study of biomembranes and lipid bilayers offers an exciting opportunity to delve into the intricate world of soft condensed matter and its intersection with the mysteries of life itself.

Protein Folding and Aggregation

Proteins are essential components of living organisms, playing crucial roles in various biological processes. They are responsible for carrying out the majority of functions within cells, including catalyzing chemical reactions, transporting molecules, providing structural support, and acting as signaling molecules. Understanding how proteins fold into their functional three-dimensional structures and how they aggregate is of utmost importance in the field of condensed matter physics.

Protein folding refers to the process by which a linear chain of amino acids, the building blocks of proteins, assumes its native, biologically active conformation. The folding process is inherently complex, as the number of possible conformations for a protein chain can be astronomically large. Proteins fold into their native structures by navigating a rugged energy landscape, where they must overcome numerous local energy minima to reach the global minimum, corresponding to their native state.

The folding process is guided by the interplay of various forces, including hydrogen bonding, van der Waals interactions, electrostatic interactions, and hydrophobic interactions. These forces determine the stability and the specific shape of the folded protein. Any disruptions or mutations in the amino acid sequence can lead to misfolding, where the protein fails to adopt its native structure. Misfolded proteins can be associated with various diseases, such as Alzheimer's, Parkinson's, and cystic fibrosis.

In addition to folding, proteins can also aggregate, forming larger structures or assemblies. Protein aggregation can occur due to various factors, including changes in temperature, pH, or the presence of denaturing agents. Aggregation can be reversible or irreversible, depending on the nature of the interactions between the protein molecules. Irreversible protein aggregation is often associated with the formation of insoluble aggregates, known as amyloid fibrils, which are implicated in neurodegenerative diseases.

Understanding the principles that govern protein folding and aggregation is not only crucial for advancing our knowledge of fundamental biological processes but also has important implications in medicine and drug design. By studying the folding and aggregation of proteins, scientists and researchers aim to develop strategies to prevent or reverse protein misfolding and aggregation, potentially leading to the development of new therapeutic approaches.

In conclusion, protein folding and aggregation are fascinating phenomena in the field of condensed matter physics. The intricate interplay of forces and interactions that govern these processes holds significant importance in understanding the fundamental principles of life. By exploring the mechanisms of protein folding and aggregation, students in the field of condensed matter physics can contribute to unraveling the mysteries of biological systems and potentially make significant contributions to medicine and healthcare.

Soft Matter in Tissue Engineering and Drug Delivery

Soft Matter in Tissue Engineering and Drug Delivery is a subchapter that explores the fascinating intersection between soft condensed matter physics and the fields of tissue engineering and drug delivery. This chapter aims to provide students with a comprehensive understanding of the fundamental principles and applications of soft matter in these rapidly evolving fields.

Tissue engineering, a multidisciplinary field, focuses on the development of functional biological substitutes to repair or replace damaged or diseased tissues. Soft matter physics plays a pivotal role in this area, as it provides insights into the mechanical and physical properties of the materials used in tissue engineering, such as hydrogels and polymers. The chapter delves into the principles of soft matter physics, including the self-assembly of molecules, viscoelasticity, and surface interactions, and how they contribute to the design and fabrication of biomaterials.

The subchapter also explores the role of soft matter in drug delivery systems. Soft materials, such as liposomes, micelles, and hydrogels, offer unique advantages in delivering therapeutic agents to specific targets in the body. The chapter discusses the principles of drug delivery, including controlled release and targeting, and how soft matter physics enables the optimization of drug delivery systems for enhanced efficacy and reduced side effects.

To provide a comprehensive understanding of the topic, the subchapter covers various experimental techniques used in soft matter physics, such as rheology, microscopy, and spectroscopy. Students will gain insights into how these techniques are employed to characterize soft materials and study their behavior in tissue engineering and drug delivery applications.

Moreover, the subchapter highlights current research trends and cutting-edge advancements in the field. Students will learn about the latest developments in tissue engineering, such as 3D bioprinting and organ-on-a-chip technologies, and how soft matter physics contributes to these advancements. The chapter also explores emerging trends in drug delivery, including the use of nanoparticles and stimuli-responsive materials.

Overall, Soft Matter in Tissue Engineering and Drug Delivery provides students in the field of condensed matter physics with a comprehensive overview of the fundamental principles and applications of soft matter in tissue engineering and drug delivery. Through this subchapter, students will gain a deeper understanding of the interdisciplinary nature of soft matter physics and its significant contributions to these rapidly evolving fields.

Chapter 6: Soft Condensed Matter in Nanotechnology

Nanomaterials and Nanoparticles

In the ever-evolving field of condensed matter physics, one area of intense interest and immense potential is the study of nanomaterials and nanoparticles. These tiny structures, ranging from 1 to 100 nanometers in size, exhibit unique properties and behaviors that differ from their bulk counterparts. As students delving into the fundamentals of soft condensed matter physics, understanding the intricacies of nanomaterials and nanoparticles is essential.

Nanomaterials, such as nanoparticles, nanowires, and nanotubes, can be synthesized from a wide range of materials, including metals, semiconductors, and polymers. The synthesis methods can vary, including top-down approaches like lithography and bottom-up techniques such as chemical vapor deposition and self-assembly. These materials possess an exceptionally high surface-to-volume ratio, enabling them to exhibit novel properties that make them highly attractive for various applications.

One of the most fascinating aspects of nanomaterials is their size-dependent properties. As the size of a material decreases to the nanoscale, quantum effects become more significant, leading to altered electronic, optical, magnetic, and thermal properties. For instance, nanoparticles of gold or silver can exhibit vibrant colors due to their surface plasmon resonance, a phenomenon resulting from their collective oscillation of conduction electrons. This unique optical property finds applications in sensing, imaging, and even cancer treatment.

Moreover, the small size of nanomaterials allows for a high degree of control over their surface chemistry and functionalization. Surface modifications, such as attaching specific molecules or functional groups, can tailor their properties for desired applications. This has paved the way for advancements in fields like drug delivery, catalysis, and energy storage.

Despite the immense potential of nanomaterials, there are also challenges associated with their synthesis, characterization, and safety. As students, it is crucial to understand the techniques employed to synthesize and manipulate nanomaterials, as well as the characterization methods used to probe their structure and properties. Furthermore, considerations surrounding the environmental and health impacts of nanomaterials are essential in their responsible development and application.

By exploring the fundamentals of nanomaterials and nanoparticles, students can gain a deeper understanding of the unique physics and chemistry at the nanoscale. This knowledge equips them with the tools to contribute to the advancement of condensed matter physics while unlocking the potential for innovative applications in various industries. As we delve further into this subchapter, we will explore specific examples, experimental techniques, and emerging trends in the exciting world of nanomaterials and nanoparticles.

Soft Condensed Matter Systems for Electronics

In recent years, there has been a growing interest in the application of soft condensed matter systems in the field of electronics. Soft condensed matter refers to materials that possess properties between those of a liquid and a solid, such as polymers, colloids, liquid crystals, and gels. These materials have unique characteristics that make them ideal for various electronic applications.

One of the most promising areas where soft condensed matter systems are being explored is in flexible electronics. Unlike traditional rigid electronic devices, flexible electronics utilize materials that can bend and stretch, enabling the creation of devices that can conform to curved surfaces or be worn on the body. Soft materials like polymers and gels are particularly attractive for this purpose due to their ability to withstand repeated mechanical deformations without losing their electrical properties.

Liquid crystals, another soft condensed matter system, have also found applications in electronics. Liquid crystal displays (LCDs) have become ubiquitous in modern televisions, smartphones, and computer monitors. The unique optical properties of liquid crystals, such as their ability to change their molecular orientation in response to an applied electric field, make them perfect for creating high-resolution displays.

Colloidal systems, which consist of tiny particles suspended in a liquid or gas, are also being explored for their potential in electronics. Colloidal materials can self-assemble into ordered structures, making them useful for creating photonic devices, such as photonic crystals or color filters. Additionally, colloidal systems can be engineered to exhibit unique electrical properties, such as tunable conductivity or higher charge storage capacity, making them attractive for energy storage applications.

Furthermore, soft condensed matter systems are being investigated for their role in energy harvesting and conversion. For instance, organic photovoltaic devices, which utilize organic semiconducting materials, have the potential to revolutionize solar energy production due to their lightweight, flexible, and low-cost nature. Similarly, thermoelectric materials, which can convert waste heat into usable electricity, can be designed using soft condensed matter principles.

In conclusion, soft condensed matter systems offer great potential for the field of electronics. Their unique properties, such as flexibility, self-assembly, and tunability, make them attractive for a wide range of applications, including flexible electronics, displays, energy harvesting, and conversion. As students of condensed matter physics, it is essential to explore and understand the fundamentals of soft condensed matter systems to contribute to the advancement of these exciting fields.

Soft Matter in Energy Applications

Soft Matter in Energy Applications is a fascinating subchapter that delves into the intersection of soft condensed matter physics and energy technologies. This topic explores the fundamental principles and applications of soft materials in various energy-related fields, providing students with a comprehensive understanding of the subject.

Soft matter refers to a class of materials that possess unique properties due to their molecular structure and intermolecular interactions. These materials include polymers, gels, liquid crystals, and colloids, which are widely used in energy applications for their versatility and functional properties.

This subchapter starts by introducing the concept of soft matter and its significance in energy technologies. It covers the basic principles of soft condensed matter physics, including the study of self-assembly, phase transitions, and rheology. Students will learn how the molecular structure and interactions of soft materials influence their behavior and performance in energy applications.

The subchapter then explores the various energy technologies where soft matter plays a crucial role. One such area is solar energy, where soft materials are used in the development of efficient solar cells and photovoltaic devices. Students will gain insights into the mechanisms of energy conversion and the design principles behind soft materials-based solar cells.

Another exciting field covered in this subchapter is energy storage. Soft matter materials, such as polymers and gels, are utilized in the development of advanced batteries and supercapacitors. Students will learn about the electrochemical processes involved in energy storage and the role of soft matter in enhancing the performance and stability of these devices.

Furthermore, the subchapter explores soft matter applications in fuel cells, thermoelectric materials, and energy harvesting devices. It discusses the unique properties of soft materials that make them suitable for these applications and the challenges faced in their development.

To enhance the learning experience, this subchapter includes numerous examples, case studies, and experimental techniques used in the study of soft matter in energy applications. Students will also find a range of exercises and problem-solving questions to test their understanding and reinforce key concepts.

Overall, Soft Matter in Energy Applications is a valuable subchapter that bridges the gap between soft condensed matter physics and energy technologies. It equips students with the knowledge and tools to explore the exciting world of soft matter and its role in shaping the future of energy.

Soft Matter in Nanomedicine and Biotechnology

Soft matter plays a crucial role in the fields of nanomedicine and biotechnology, where the study and manipulation of materials at the nanoscale have revolutionized the way we approach medical treatments and biological systems. In this subchapter, we will explore the fundamentals of soft condensed matter physics in the context of nanomedicine and biotechnology, providing students with a comprehensive understanding of this exciting field.

Soft matter refers to a wide range of materials that have properties between those of conventional liquids and solids. It includes polymers, colloids, liquid crystals, and biological macromolecules such as proteins and DNA. These materials are highly flexible, responsive, and often exhibit self-assembly properties, making them ideal candidates for various applications in nanomedicine and biotechnology.

One of the key areas where soft matter has made significant contributions is in drug delivery systems. Soft nanoparticles, such as liposomes and polymeric micelles, can encapsulate therapeutic agents and efficiently transport them to specific sites in the body. Their high stability, biocompatibility, and ability to release drugs in a controlled manner have revolutionized the field of targeted drug delivery, enabling more effective treatments with reduced side effects.

Furthermore, soft matter physics plays a vital role in understanding and engineering biomaterials for tissue engineering and regenerative medicine. By mimicking the extracellular matrix, researchers can create scaffolds that promote cell adhesion, proliferation, and differentiation, facilitating the growth of new tissues and organs. Soft materials also enable the development of biosensors, diagnostic devices, and smart materials that respond to external stimuli, providing real-time monitoring and personalized medicine solutions.

In this subchapter, students will delve into the fascinating properties of soft matter at the nanoscale, such as self-assembly, phase transitions, and rheology, and learn how these properties can be harnessed for applications in nanomedicine and biotechnology. They will explore the interplay between physics, chemistry, and biology in the design and fabrication of soft materials and gain an understanding of the challenges and opportunities in this interdisciplinary field.

By studying soft matter in the context of nanomedicine and biotechnology, students will not only gain a deeper understanding of condensed matter physics but also appreciate the potential of soft materials in revolutionizing healthcare and biological systems. This subchapter aims to inspire students to explore further research and innovation in this exciting field, contributing to the advancement of nanomedicine and biotechnology for the benefit of society.

Chapter 7: Experimental Techniques in Soft Condensed Matter Physics

Optical Microscopy and Spectroscopy

In the field of Condensed Matter Physics, the study of soft condensed matter is of utmost importance. Soft condensed matter refers to materials that possess properties between those of a solid and a liquid, such as polymers, colloids, and liquid crystals. Understanding the fundamental principles behind these materials is crucial for advancements in various fields, including material science, biology, and nanotechnology.

When it comes to investigating soft condensed matter, optical microscopy and spectroscopy techniques play a vital role. These techniques allow scientists to observe and analyze the structure, dynamics, and optical properties of materials at the micro- and nano-scale.

Optical microscopy involves using visible light to magnify and image samples. It provides valuable information about the morphology and topography of soft condensed matter systems. By utilizing different microscopy techniques, such as bright-field, dark-field, and fluorescence microscopy, researchers can visualize the intricate details of these materials in real-time. Moreover, the advancements in imaging technologies have enabled the development of super-resolution microscopy, which allows scientists to surpass the diffraction limit and observe structures with nanometer resolution.

On the other hand, spectroscopy techniques provide insights into the optical properties and chemical composition of soft condensed matter. By analyzing the interaction between light and matter, spectroscopy helps to identify and quantify various characteristics such as absorption, emission, and scattering. Techniques like UV-Vis spectroscopy, Fourier-transform infrared spectroscopy (FTIR), and Raman spectroscopy are commonly employed to study soft condensed matter systems. These techniques enable researchers to investigate the electronic and vibrational properties of materials, which are crucial for understanding their behavior and functionality.

The combination of optical microscopy and spectroscopy offers a powerful toolset for studying soft condensed matter physics. By integrating these techniques, scientists can simultaneously observe the structure and dynamics of materials while analyzing their optical properties. This multidimensional approach provides a comprehensive understanding of soft condensed matter systems and enables the design and development of novel materials with tailored properties.

As students exploring the fundamentals of soft condensed matter physics, it is crucial to familiarize ourselves with optical microscopy and spectroscopy techniques. These tools not only allow us to visualize and characterize materials but also provide a platform for innovation and discovery. By mastering these techniques, we can contribute to the advancement of various fields and become proficient researchers in the exciting realm of soft condensed matter physics.

X-ray and Neutron Scattering

X-ray and Neutron Scattering: Probing the Secrets of Soft Condensed Matter

In the fascinating world of soft condensed matter physics, the tools of X-ray and neutron scattering have emerged as powerful techniques for unraveling the intricate structure and dynamics of materials at the atomic and molecular level. This subchapter explores the fundamental principles and applications of X-ray and neutron scattering, shedding light on how these techniques enable us to peer into the hidden secrets of soft condensed matter.

X-ray and neutron scattering involve the interaction of X-rays and neutrons with matter, leading to the scattering of these particles in various directions. By analyzing the scattered radiation, we can gain valuable insights into the structure and dynamics of soft condensed matter systems. These techniques have revolutionized our understanding of a wide range of materials, from biological macromolecules to polymers and nanoparticles.

One of the key advantages of X-ray and neutron scattering is their ability to probe the atomic and molecular structure of materials without destroying or altering them. This non-invasive nature allows for in-situ measurements, making it possible to study dynamic processes and phase transitions in real-time. Moreover, X-ray and neutron scattering can provide information about the size, shape, and orientation of nanoparticles and macromolecules, helping us understand how these entities interact and self-assemble.

The subchapter delves into the underlying principles of X-ray and neutron scattering, explaining the differences between the two techniques and their complementary nature. It discusses the basics of wave-particle duality, diffraction, and scattering theory, providing a solid foundation for understanding the experimental methods used in soft condensed matter physics.

Furthermore, the subchapter highlights the diverse applications of X-ray and neutron scattering in soft condensed matter research. From elucidating the structure of biological molecules such as proteins and DNA to studying the behavior of liquid crystals and colloidal suspensions, these techniques have contributed immensely to our understanding of complex materials. The subchapter also explores recent advancements in the field, including the development of novel scattering techniques and the use of synchrotron and neutron sources.

Overall, this subchapter on X-ray and neutron scattering serves as a gateway for students into the captivating world of soft condensed matter physics. By providing an in-depth understanding of how these techniques work and their applications, it equips students with the knowledge and tools necessary to explore the mysteries of complex materials and contribute to advancements in the field of condensed matter physics.

Electron Microscopy and Scanning Probe Microscopy

In the fascinating world of condensed matter physics, the study of soft materials holds a crucial place. To comprehend the intricacies of soft condensed matter, scientists rely on advanced microscopic techniques that offer unparalleled insights into the structure and behavior of these materials. Two such remarkable techniques are electron microscopy and scanning probe microscopy.

Electron microscopy, as the name suggests, involves the use of electrons to image and analyze samples at incredibly high resolutions. Unlike traditional light microscopes limited by the diffraction of light waves, electron microscopes exploit the wave-particle duality of electrons to achieve resolutions as low as the atomic scale. This makes electron microscopy an indispensable tool in understanding the complex structure and dynamics of soft condensed matter.

There are two primary types of electron microscopy: transmission electron microscopy (TEM) and scanning electron microscopy (SEM). TEM involves passing a beam of electrons through a thin sample, allowing researchers to observe its internal structure in great detail. This technique is particularly useful for studying the arrangement of atoms, defects, and interfaces within soft materials. On the other hand, SEM provides a detailed surface view of samples by scanning a focused electron beam across the surface and detecting the emitted electrons. SEM is often employed to investigate the topography, composition, and morphology of soft materials.

Another powerful microscopy technique used in condensed matter physics is scanning probe microscopy (SPM). SPM operates by scanning a sharp probe over the surface of a sample to create a detailed topographic map. The probe interacts with the material, generating precise measurements of various properties, such as electrical conductivity, magnetic fields, and even chemical composition. The most common types of SPM are atomic force microscopy (AFM) and scanning tunneling microscopy (STM). AFM measures the forces between the probe and the sample, while STM uses quantum tunneling phenomena to detect the variations in electron density. SPM techniques provide valuable information about the surface properties and mechanical behavior of soft condensed matter.

The marriage of electron microscopy and scanning probe microscopy has revolutionized the field of condensed matter physics, enabling scientists to explore the fundamental aspects of soft materials like never before. By visualizing and manipulating matter at such small scales, researchers are unraveling the mysteries behind the unique properties and functionalities exhibited by soft condensed matter. These techniques open up new avenues for designing and engineering innovative materials with applications ranging from electronics and energy storage to biomedical sciences.

As students of condensed matter physics, understanding the principles and applications of electron microscopy and scanning probe microscopy will equip you with vital tools for your scientific journey. These techniques will enable you to explore the intricacies of soft condensed matter, unravel its underlying principles, and contribute to the development of cutting-edge technologies that shape our world.

Rheological and Mechanical Characterization Techniques

Soft condensed matter physics is a fascinating field that deals with the study of materials with unique properties, such as polymers, colloids, and liquid crystals. Understanding the behavior of these materials requires a deep knowledge of their rheological and mechanical properties. In this subchapter, we will explore the various techniques used to characterize these properties and how they contribute to our understanding of soft condensed matter physics.

Rheology is the study of how materials flow and deform under the influence of an applied force. It plays a crucial role in understanding the behavior of soft materials, as it provides insights into their viscoelastic properties. One commonly used technique in rheology is shear rheometry, which measures the response of a material to an applied shear stress. This technique allows us to determine the material's viscosity, elasticity, and complex modulus, providing valuable information about its flow behavior.

Another important technique in characterizing soft condensed matter is mechanical testing. This involves subjecting materials to different mechanical forces, such as tension, compression, or bending, and studying their response. For example, the tensile test measures the material's response to stretching forces, providing information about its strength and elasticity. Similarly, the compression test reveals the material's ability to withstand compressive forces.

To accurately characterize the mechanical properties of soft materials, it is essential to use specialized equipment. For instance, atomic force microscopy (AFM) allows us to probe the mechanical properties of materials at the nanoscale. AFM uses a tiny probe to measure forces between the probe and the material's surface, providing information about its mechanical properties with high spatial resolution.

In addition to rheological and mechanical characterization techniques, other methods, such as microscopy and spectroscopy, are also employed in soft condensed matter physics. Microscopy techniques, including optical microscopy and electron microscopy, enable the visualization of material structures and the observation of dynamic processes. Spectroscopy techniques, on the other hand, provide insights into the material's molecular and electronic properties, aiding in understanding its behavior.

In conclusion, the rheological and mechanical characterization techniques discussed in this subchapter are vital tools in understanding the fundamentals of soft condensed matter physics. By employing these techniques, scientists can gain valuable insights into the flow behavior, viscoelastic properties, and mechanical response of soft materials. These findings are essential for advancements in various fields, including materials science, biophysics, and nanotechnology. As students of condensed matter physics, gaining a thorough understanding of these techniques will enable us to delve deeper into the fascinating world of soft condensed matter.

Chapter 8: Applications of Soft Condensed Matter Physics

Soft Condensed Matter Physics in Materials Science

Soft condensed matter physics plays a crucial role in the understanding and development of materials science. This subchapter aims to provide students with an insightful perspective on the intersection of soft condensed matter physics and materials science, highlighting its significance in the field of condensed matter physics.

Materials science deals with the study of the properties and behaviors of various materials, ranging from metals and ceramics to polymers and colloids. Soft condensed matter physics focuses on understanding the physical properties of materials that are easily deformable, such as liquids, gels, foams, and biological systems. This subchapter discusses how the principles and concepts of soft condensed matter physics contribute to the understanding and manipulation of these materials.

One of the key aspects of soft condensed matter physics in materials science is the understanding of the structure and dynamics of materials at the molecular and mesoscopic levels. Soft materials often exhibit complex hierarchical structures, where the arrangement of molecules or particles at different length scales governs their properties. By studying the interactions between these constituents and the resulting emergent behavior, researchers can gain insights into the design and control of material properties.

Moreover, the subchapter explores the role of soft condensed matter physics in understanding the mechanical properties of materials. Soft materials often possess unique mechanical responses due to their ability to deform and flow under external forces. Understanding the underlying physics behind this behavior is crucial for designing materials with desired mechanical properties, such as elasticity, toughness, and strength. The subchapter also discusses the application of soft condensed matter physics principles in the development of smart materials, which can respond to external stimuli, such as temperature, light, or electric fields.

Furthermore, the subchapter delves into the importance of soft condensed matter physics in the field of biological materials. Biological systems, such as cell membranes, proteins, and DNA, exhibit fascinating soft matter behaviors. By applying the principles of soft condensed matter physics, researchers can better understand the structure and function of biological materials, leading to advancements in drug delivery systems, tissue engineering, and biomaterials.

In conclusion, this subchapter provides students with a comprehensive overview of the role of soft condensed matter physics in materials science. By understanding the principles and concepts discussed, students can appreciate the significance of soft condensed matter physics in the study, design, and development of materials with tailored properties. Whether one's interest lies in condensed matter physics or materials science, this subchapter offers a valuable perspective for students aspiring to delve into these niches.

Soft Matter in Food Science and Cosmetics

Soft matter plays a crucial role in various fields, including food science and cosmetics. In this subchapter, we will delve into the fascinating world of soft matter and explore its applications in these industries. Designed for students and enthusiasts of condensed matter physics, this section will provide a comprehensive overview of the fundamentals and practical aspects of soft matter in food science and cosmetics.

Soft matter refers to a class of materials that possess unique properties, such as deformability and sensitivity to environmental conditions. These materials include polymers, colloids, gels, foams, and liquid crystals, among others. In food science, soft matter is a key component in understanding the behavior and structure of complex food systems. From understanding the emulsion stability in mayonnaise to the rheology of ice cream, soft matter principles are essential in developing and improving food products.

One of the key areas where soft matter physics finds application in food science is in the study of colloidal systems. These systems consist of small particles dispersed in a liquid medium, such as milk or fruit juices. By understanding the interactions between colloidal particles, scientists can control the stability, texture, and sensory properties of food products. For example, the addition of stabilizers like hydrocolloids can enhance the shelf life and mouthfeel of processed foods.

In the field of cosmetics, soft matter physics also plays a crucial role. Cosmetics are composed of complex mixtures of emulsions, suspensions, and gels, which require a deep understanding of soft matter principles for formulation and stability. The structure and rheology of creams, lotions, and shampoos are influenced by the interactions between various soft matter components. By studying these interactions, scientists can design products with desired properties, such as enhanced spreadability, stability, and release.

Furthermore, soft matter physics also contributes to advancements in drug delivery systems and biomaterials used in the medical field. By understanding the behavior of soft matter materials in biological environments, researchers can develop innovative drug carriers and scaffolds for tissue engineering.

In conclusion, soft matter physics plays a crucial role in various industries, including food science, cosmetics, and medicine. By studying and understanding the principles of soft matter, students in the field of condensed matter physics can contribute to the development of improved food products, cosmetics, and biomaterials. This subchapter aims to provide a solid foundation in soft matter physics, specifically in relation to food science and cosmetics, allowing students to explore the exciting potential of this field.

Soft Matter in Environmental Science

Soft Matter in Environmental Science is a subchapter in the book "Exploring the Fundamentals of Soft Condensed Matter Physics: A Student's Perspective," specifically addressing students with a keen interest in Condensed Matter Physics. This chapter delves into the fascinating world of soft matter and its applications in environmental science.

Soft matter refers to materials that possess properties between those of conventional solids and liquids. These materials are ubiquitous in our environment and play a crucial role in various natural and man-made systems. Understanding the fundamental properties of soft matter is essential for comprehending their behavior and designing innovative solutions for environmental challenges.

One significant aspect of soft matter in environmental science is its impact on pollution and waste management. For instance, the chapter explores how polymers, a class of soft matter, are used in the development of eco-friendly materials. These materials can replace single-use plastics and reduce environmental degradation caused by non-biodegradable waste. Furthermore, the chapter discusses the application of soft matter in smart materials for pollution sensing and remediation. These materials can detect and absorb pollutants, leading to cleaner air and water.

Another area of focus is the role of soft matter in climate change mitigation and adaptation. The chapter explains how soft matter researchers are developing novel materials for energy storage and conversion. These materials offer efficient and sustainable alternatives to traditional energy sources, such as fossil fuels. Additionally, the chapter explores how soft matter can be used in the design of resilient and eco-friendly infrastructure to combat the effects of climate change, such as rising sea levels and extreme weather events.

Furthermore, the subchapter delves into the study of soft matter interfaces in environmental science. These interfaces are crucial in understanding the behavior of pollutants, the transport of contaminants in soils and water, and the interactions between soft matter and biological systems. The chapter provides insights into the physics governing these interfaces and their implications for environmental processes.

In conclusion, the subchapter on Soft Matter in Environmental Science offers students interested in Condensed Matter Physics a comprehensive overview of the applications and significance of soft matter in addressing environmental challenges. By understanding the properties and behavior of soft matter materials, students can contribute to the development of sustainable solutions for pollution control, climate change mitigation, and the preservation of our environment.

Future Perspectives and Challenges in Soft Condensed Matter Physics

Soft condensed matter physics is a fascinating field that studies the behavior of materials that are neither solid nor liquid, such as polymers, colloids, liquid crystals, and biological macromolecules. As students delving into the world of condensed matter physics, it is essential to understand the future perspectives and challenges that lie ahead in this dynamic field.

One of the most exciting future perspectives in soft condensed matter physics is the development of new materials with tailored properties. Scientists are actively working to design and synthesize materials with unique characteristics for applications in energy storage, drug delivery systems, and advanced electronics. By understanding the fundamental principles governing the behavior of soft matter, students can contribute to the creation of innovative materials that can revolutionize various industries.

Another promising avenue in soft condensed matter physics is the exploration of self-assembly and self-organization phenomena. Nature has perfected self-assembly processes over billions of years, leading to the formation of complex structures with remarkable functionalities. By studying the principles behind self-assembly, students can contribute to the development of novel materials that can organize themselves into desired structures, opening up possibilities for advanced nanotechnology and biomaterials.

Furthermore, the interdisciplinary nature of soft condensed matter physics presents both opportunities and challenges. The field requires a deep understanding of physics, chemistry, and biology, as well as expertise in experimental techniques and theoretical modeling. As students, it is crucial to embrace this interdisciplinary approach, collaborate with researchers from different backgrounds, and acquire knowledge in diverse fields to tackle complex problems at the interface of different disciplines.

However, soft condensed matter physics also faces several challenges. One of the main challenges is understanding the dynamics of complex systems. Soft matter often exhibits non-equilibrium behavior, making it difficult to predict and control its properties. Developing theoretical frameworks and experimental techniques that can capture the dynamics of soft matter at different length and time scales is a major hurdle that needs to be overcome.

Another challenge is the integration of soft matter physics with emerging technologies. As new technologies such as artificial intelligence, quantum computing, and nanotechnology continue to advance, it is essential to explore how soft matter can contribute to and benefit from these developments. This requires students to stay updated with the latest advancements in relevant fields and actively seek collaborations and interdisciplinary research opportunities.

In conclusion, the future perspectives of soft condensed matter physics are incredibly promising, with opportunities for designing new materials, understanding self-assembly processes, and exploring interdisciplinary research. However, challenges such as understanding complex dynamics and integrating soft matter with emerging technologies need to be addressed. As students, embracing interdisciplinary approaches and staying informed about the latest advancements will be crucial in contributing to the exciting future of soft condensed matter physics.

Chapter 9: Conclusion and Further Reading

Summary of Key Concepts

In this subchapter, we will provide a comprehensive summary of the key concepts discussed throughout the book, "Exploring the Fundamentals of Soft Condensed Matter Physics: A Student's Perspective." Designed specifically for students interested in the field of Condensed Matter Physics, this summary aims to consolidate your understanding of the fundamental principles and essential concepts covered in the book.

1. Soft Condensed Matter: We begin by defining soft condensed matter as a branch of physics that focuses on the study of materials with properties between those of rigid solids and fluids. This includes polymers, colloids, liquid crystals, and biological materials.

2. Molecular Interactions: Understanding the interactions between molecules is crucial in soft condensed matter physics. We explore various intermolecular forces, including van der Waals forces, electrostatic interactions, hydrogen bonding, and hydrophobic interactions.

3. Phase Transitions: Phase transitions occur when a substance undergoes a change in its physical state. We delve into different types of phase transitions, such as the liquid-gas transition, ordering transitions in liquid crystals, and gelation transitions in polymers.

4. Rheology: Rheology is the study of how materials deform and flow under applied stress. We discuss the basic principles of rheology,

including stress, strain, viscosity, and viscoelasticity, and how they apply to soft condensed matter systems.

5. Self-Assembly: Self-assembly is a fascinating phenomenon where molecules or particles spontaneously organize themselves into ordered structures. We explore the principles of self-assembly and its applications in soft condensed matter, such as the formation of micelles, vesicles, and colloidal crystals.

6. Biological Soft Matter: Soft condensed matter physics plays a significant role in understanding biological systems. We delve into the physics of biological materials, including proteins, DNA, and cell membranes, and how their properties relate to their functions.

7. Active Matter: Active matter refers to systems composed of self-propelled particles that exhibit collective behavior. We discuss the physics behind active matter, including the emergence of flocking, swarming, and self-organization.

8. Experimental Techniques: We provide an overview of experimental techniques commonly used in soft condensed matter physics, such as X-ray scattering, microscopy, spectroscopy, and rheology measurements.

By summarizing these key concepts, this subchapter aims to reinforce your understanding of the core principles of soft condensed matter physics. Whether you are a beginner or an advanced student in the field, this summary will serve as a valuable reference, enabling you to navigate the fascinating world of soft condensed matter with confidence.

Recommended Books and Resources for Further Study

As students delving into the fascinating field of Condensed Matter Physics, it is crucial to have a solid foundation of knowledge and resources at your disposal. This subchapter aims to provide you with a curated list of recommended books and resources that will not only deepen your understanding of soft condensed matter physics but also serve as invaluable references throughout your academic journey.

1. "Soft Condensed Matter" by Richard A.L. Jones: This comprehensive textbook offers a clear and concise introduction to the fundamentals of soft condensed matter physics. It covers a wide range of topics, including liquid crystals, polymers, colloids, and biological materials, making it an essential read for beginners.

2. "Introduction to Solid State Physics" by Charles Kittel: While primarily focused on solid-state physics, this classic textbook lays a solid foundation for understanding the principles that govern condensed matter physics. It explores the structure, properties, and behavior of various condensed matter systems, providing a comprehensive overview of the subject.

3. "The Physics of Soft and Biological Matter" by A. Fernandez-Nieves and Antonio Fernandez-Barbero: This book adopts a multidisciplinary approach, bridging the gap between soft matter physics and biology. It explores the physics of biological materials, such as membranes and cells, as well as the behavior of soft matter systems, including gels and liquid crystals. It is an excellent resource for students interested in the intersection of biology and condensed matter physics.

4. "Statistical Mechanics of Membranes and Surfaces" by David Nelson: This text focuses on the statistical mechanics of thin elastic sheets, such as biological membranes and fluid interfaces. It provides a comprehensive introduction to the theoretical aspects of these systems, covering topics such as bending elasticity, phase transitions, and fluctuations in membranes. This book is highly recommended for those interested in the mechanical properties of soft matter.

In addition to these books, there are several online resources that can greatly enhance your understanding of soft condensed matter physics. Websites such as the Soft Matter World and the Soft Matter Journal provide access to a wealth of research papers, reviews, and educational materials. Online lecture series, such as those offered by the Perimeter Institute and the MIT OpenCourseWare, can also be invaluable resources for further study.

By utilizing these recommended books and resources, you will have a strong foundation in soft condensed matter physics and be well-equipped to delve deeper into this exciting field. Remember to supplement your studies with hands-on experimentation and collaboration with peers and mentors to truly grasp the intricacies of soft condensed matter physics.

Future Directions in Soft Condensed Matter Physics Research

As students of Condensed Matter Physics, it is essential to be aware of the ever-evolving nature of research in this field. Soft Condensed Matter Physics, in particular, has witnessed tremendous growth and offers exciting opportunities for future exploration. In this subchapter, we will discuss some of the potential future directions that researchers may undertake in this captivating area of study.

One promising avenue of research lies in the field of self-assembly and self-organization. Soft materials, such as polymers, colloids, and liquid crystals, have the remarkable ability to spontaneously arrange themselves into intricate and ordered structures. Understanding the fundamental principles that govern self-assembly and harnessing this phenomenon holds immense potential for various applications, including drug delivery systems, advanced materials design, and nanotechnology.

Another area of interest is active matter. Active matter refers to systems composed of individual units that consume energy and display collective motion. Examples of active matter include swarms of bacteria, schools of fish, and even artificial micro-robots. Investigating the emergent behaviors and the underlying principles of active matter can provide insights into biological processes, enhance our understanding of complex systems, and inspire the development of novel bio-inspired technologies.

The study of soft matter interfaces also offers intriguing possibilities for future research. Interfaces are the boundaries between different phases or materials, and they often exhibit unique properties and behaviors. Exploring the dynamics and interfacial phenomena in soft matter systems can lead to advancements in areas such as surface coatings, microfluidics, and biomaterials. Additionally, understanding the interplay between interfaces and external stimuli, such as light, electric fields, or temperature, can open up avenues for developing responsive and adaptive soft materials.

Furthermore, the combination of soft matter with other disciplines, such as biology, chemistry, and materials science, presents exciting interdisciplinary research opportunities. Collaborations between scientists from various fields can lead to breakthroughs in areas such as tissue engineering, drug delivery, and sustainable energy. By combining expertise and perspectives from different disciplines, researchers can tackle complex problems and develop innovative solutions.

In conclusion, the future of Soft Condensed Matter Physics research is filled with immense possibilities. From self-assembly and active matter to interfaces and interdisciplinary collaborations, the field offers a wide range of avenues for exploration. As students, it is crucial to stay abreast of these future directions and be prepared to contribute to the ever-expanding knowledge in this fascinating field. By embracing these challenges and pursuing research in these frontier areas, we can shape the future of Soft Condensed Matter Physics and make significant contributions to science and technology.